Riparia's River

Michael J. Caduto   Illustrated by Olga Pastuchiv

It was a sultry summer day, the kind that would make a frog sweat
if it could. Gretchen and her twin brother, Jason, led their
younger sister and Jason's best friend, Mark, along the riverbank
to their favorite swimming hole.

TILBURY HOUSE, PUBLISHERS • GARDINER, MAINE

Race you to the rope swing!" Jason shouted to Mark. They took off along the narrow path, trying to pass each other and be first to reach the rope on the huge oak that branched out above the water.

"I beat you!" said Jason as he swung over the deep pool in the bend of the river.

At the farthest reach of the rope, Jason let go and dropped into the water. The instant his head came up, Mark landed a perfect cannonball beside him.

When Mark surfaced, Jason was coughing and pushing aside slime. "What is wrong with this water? It smells like a sewer," he said.

"You're not kidding, and what is this?" Mark asked, lifting his hand covered with green stuff.

Both boys scrambled out of the river.

Gretchen caught up with them, sniffing the air. "What did you two do in that water?"

"Very funny," Jason said.

Their younger sister Daphne trundled up the trail. "You never wait for me!"

"Well, we're all here now," said Gretchen. "But I'm not going to swim in *that*."

"Me neither," said Daphne, holding her nose.

"Maybe if we go upriver we can see what's going on," Gretchen said.

"Why up the river?" asked Daphne.

"Water flows downhill, so we need to look *up* the river to see where this slime is coming from."

After walking for a while, Gretchen spotted someone else on the riverbank. A woman stood surrounded by the arching branches of an ancient tree.

"Hello. Are you looking for someone?"

"Not exactly," said Gretchen warily.

"I am Riparia," she said, smiling.

Daphne repeated the name to herself in a slow whisper, as she did with any big new word. "Rye-PEAR-ee-ah?"

"Yes," Riparia said briskly. "My name means 'of the riverbank.'"

She looked at Mark and Jason. "I see you've been swimming. You shouldn't go in the river right now. The river isn't well."

"Yeah," said Jason. "We kind of figured that out the hard way. Do you know what happened to the water?"

"Come with me," said Riparia, as she started to walk upriver. "I'll show you."

At first, the children looked around, silently asking each other if they should follow Riparia. Then Jason ran after her and called, "Wait for us!"

Walking behind Riparia, the children saw that she moved as gracefully as a deer. Her feet touched the ground so gently that they hardly made a sound.

The bank dipped down to the edge of a grove of river birches. Nearby, the passing water sang over rocks and riffles.

"See how the trees cast shade over the water," said Riparia.

"Uh, sure," Mark replied.

"That's what keeps the water cool. Trout and other kinds of fish, and the insects they eat, need cool, clean water."

"Why?" asked Gretchen.

"Cool water has more of the oxygen that fish breathe through their gills. The cooler, the better. Insects like it, too. But the green stuff you found in your swimming hole uses up a lot of oxygen."

"I don't see any insects," said Jason.

"No, you wouldn't see them unless you looked under the rocks, or if you came and looked at night when many insects crawl up onto the tops of rocks to feed," said Riparia.

Riparia led them farther upstream. The lush tangle of riverside plants gave way to the edge of a cornfield, which went right down to the water.

"This is Amy's family farm," Gretchen said.

Riparia pointed to the eroded riverbank. "The farmer sprayed a weed killer on the cornfield. When it rains, the chemicals run downhill and into the water. See how the dirt is washed away?"

"But Amy's dad needs corn to feed his cows," Gretchen said.

"You're right. The farmer can plant his corn, just not all the way down to the water."

"How far down?" asked Mark.

"Do you have a good throwing arm?" asked Riparia.

Mark puffed out his chest, held up his arm to flex his bicep, and said, "Look at that! You bet I do."

"If you stood at the edge of the water and threw a stone as far as you could into the cornfield, it would land about as close to the river as the farmer should plant in order to protect the water."

She eyed Mark's bicep. "That would be about one hundred feet."

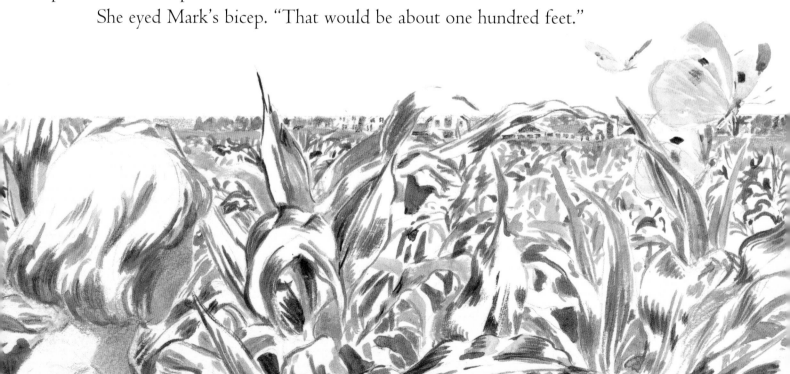

That land between the water and the cornfield is called a buffer. The plants that grow naturally within the buffer keep the water clean," Riparia said.

"How?" asked Gretchen, crouching down to get a closer look.

"When rainwater runs down to the river, it carries topsoil, fertilizer, and chemicals with it. Plants slow the water down so that most of the topsoil settles out. Roots act like fingers that grip the soil and hold it in place. Plus, the leaf litter and roots absorb lots of chemicals and fertilizers. A buffer helps clean the runoff before it reaches the river."

Jason yawned and looked away. "Is this where you were taking us?" he asked impatiently.

Riparia shook her head and beckoned them on. This time little Daphne was the first to follow, and the rest fell in behind her.

At the edge of the next field, a barbed-wire fence curved down into the river. Cows grazed there and wandered into the water to drink and cool off. The soil was muddy and pocked with hoof prints.

Gretchen sniffed the air. "That's the smell at the swimming hole! Is it coming from those cow pies?"

"Cow pies?" asked Daphne.

"You know, cow poop."

"Yes, and that's not all," said Riparia. "The manure and soil that wash into the river act like fertilizer, feeding the algae in the water."

"What's algae?" asked Mark.

"Tiny, microscopic plants. There are millions of them in the river. When too much fertilizer gets into the water the algae grow quickly. That's how the slime got into your swimming hole. The slime is algae fed by the cow manure and fertilizer from the cornfield. Then, when the plants die and decay, the oxygen gets used up in the river and—"

"—and those bugs and trout can't live there," Gretchen said.

"Exactly," Riparia nodded. "Plus, the decaying algae smells like rotten eggs."

’m friends with Amy, the farmer’s daughter,” said Gretchen. “We’re going to have a little chat about all this. Her dad’s farm is hurting the river!”

“And ruining our swimming hole,” Jason added.

“Careful,” Riparia cautioned. “You don’t want to upset your friend. No one likes to be told what to do.”

“Well, then what *can* we do?” asked Jason. “I’m not going to sit here and watch our swimming hole go down the—well—down the river.”

“What if we invited Amy to come swim with us?” Mark asked.

Riparia smiled, “Good idea.”

For a while, they all stood and watched the cows drinking and wading along the shore.

When Gretchen turned to ask a question, Riparia was gone. “Where did she go?”

They all looked around. Riparia was nowhere to be seen.

“How did she do that?” asked Daphne, walking over to Gretchen and hugging her waist.

“That’s awesome,” said Jason. “I want Riparia to teach me how to move like that. She’s as quiet as a fox.”

It was even hotter the next day as Amy joined the group for a swim.

When she saw them standing by the rope swing, she asked, "Why isn't anyone going in?"

"Check it out," Jason said.

Amy wrinkled her nose. "What's that smell?"

"Come with us," said Gretchen, "and we'll show you."

Gretchen led the group upstream to the cornfield and the cow pasture beyond. She explained how water runs off from the farm fields and hurts the river.

"I didn't realize so much stuff was washing into the water," Amy said. "It's really gross."

"Can you talk to your father about it?" asked Gretchen.

"I can try," Amy replied, looking down at the ground. "He's really busy, and things aren't going too well with the farm, so I don't know what he'll say." She looked up at her friends. "But I'll try."

The following day, everyone sat down under the big sugar maple behind the twins' house. No one spoke. They just stared at Amy.

"My father says he doesn't have the time or money to deal with it."

"So what are we going to do?" asked Mark.

Amy continued, "He also said that if we could figure something out that wouldn't cost him any money, we could fix the problem ourselves. He said it might keep us out of trouble." She rolled her eyes.

"Hey, I know!" cried Mark. "If we do what Riparia says, and move the fence back a hundred feet, and help things grow back along the riverbank, then the water would get clean again."

"Riparia?" Amy asked.

"A woman we met by the river. She told us about something called a buffer, a stretch of land between a farm field and the river, which would help keep the river clean," Gretchen said. Then she turned to Mark. "Doing all that work would take time—lots of time and lots of energy."

"I've got time," said Mark.

"I've got energy," Amy said.

"Me, too," said Daphne as she planted her feet, folded her arms, and stuck out her chin.

Jason looked over at her like he was about to laugh. But instead he said, "Me, three," and put his arm around his fierce little sister.

During the weeks that followed, the children worked. Amy's dad helped them remove the barbed wire from the fence posts. They set the posts back away from the river, then restrung the wire.

When the fence was done, Mark and Jason slapped high fives.

"Dad," said Amy, "there's one more thing."

"I know, Amy, I know. Goodness knows you've asked enough times. Starting next spring, I'll only plant corn down to a hundred feet from the river."

Amy smiled. "Thanks, Dad."

The next morning, the children started to move plants from other parts of the riverbank to their new buffer. At the end of that hot day of hard work, Amy said, "There is way too much work here for just the five of us. We need to get more of our friends to help."

As the days passed, word spread, and more and more people arrived to help. Gretchen told the helpers everything that Riparia said about why it was important to have plants growing along the riverbank—about all of the good things the plants did for the water and for life in the river.

Throughout that summer and the next, the children transplanted hundreds
of nearby wildflowers, shrubs, and trees between the river and the new fence-line.
They planted red maple and cottonwood, green ash, willow, and sycamore. Then they
planted alder, blueberry, elderberry, groundnut, boneset, joe-pye weed, and lots of
ferns. Daphne liked it best when they put on their bathing suits and old sneakers
and mucked out into the shallow mud to plant cattails and pickerel weed.

ime passed. The plants grew. Seeds of other plants arrived, falling from the feathers and fur of visiting birds and animals, and left behind in their droppings. Within two years a beautiful stretch of wildness grew along the riverbank where the corn and muddy cow pasture used to be.

One day, the children spotted monarch butterflies flitting among the milkweed blossoms and heard a pair of goldfinches fly past singing a sweet song with

each dip of their flight: "Potato-chip, potato-chip."

"Look at that," said Jason. "The riverbank we planted has turned into a real home for these animals. A real, you know—"

"—habitat," said Gretchen.

"Right!" replied Jason.

"Riparia would be happy," Amy said.

"It looks like *someone* is," Daphne laughed as a monarch fluttered by her nose.

"Last one in is a rotten egg," exclaimed Jason as he ran toward the riverbank yelling, "Yee hah!"

The Fauna of Riparia's River
(Can you find them in this book?)

Birds
American Woodcock
Barn Swallow
Belted Kingfisher
Brown Creeper
Common Screech Owl
Eastern Meadowlark
Great Blue Heron
Hairy Woodpecker
Killdeer
Mallard Ducks
Quail (Common Bobwhite)
Red-Winged Blackbirds
White-Breasted Nuthatch
Wood Ducks
Wood Thrush

Butterflies
American Copper
Cabbage White
Compton Tortoiseshell
Great Spangled Fritillary
Meadow Fritillary
Monarch
Orange Sulfur
Pearl Crescent
Red Admiral
Swallowtail
White Admiral

Dragonflies
Comet Darner
Common Whitetail
Flame Skimmer
Gray Petaltail
Halloween Pennant
Widow Skimmer

Fish
Blacknose Dace
Brook Trout
Creek Chub
Fathead Minnow

Larvae
Blue-Winged Olive
   (Mayfly Nymph)
Dragonfly Nymph
Golden Stonefly Creeper
   (Stonefly Nymph)
Leadwing Coachman
   (Mayfly Nymph)
Northern Case-Maker
   (CaddisflyLarva)

Mammals
Chipmunk
Deer Mouse
Gray Squirrel
Raccoon
White-Tailed Deer

Reptiles & Amphibians
Gray Treefrog
Red-Bellied Turtles (Juveniles)
Salamander

Others
Carolina Grasshopper
Common Meadow Katydid
Ladybug (Ladybird Beetle)
Praying Mantis
Snail

TILBURY HOUSE, PUBLISHERS
103 Brunswick Avenue, Gardiner, Maine 04345
800–582–1899 · www.tilburyhouse.com

First hardcover edition: April 2011 · 10 9 8 7 6 5 4 3 2

To every child who has ever experienced a sense of joy, of wonder and awe, while discovering the mysteries of nature at the water's edge.
For a particular water-loving child who grew up to become a watershed coordinator for the State of Vermont, a passionate protector of all things aquatic, and my wife—Marie Levesque Caduto. —MJC

For family and friends, especially Ivee and Kiva, Helen and Emily, and Michael, Lauren, and Samuel. —OP

Library of Congress Cataloging-in-Publication Data

Caduto, Michael J.
Riparia's river / Michael J. Caduto ; illustrated by Olga Pastuchiv. — 1st hardcover ed.
    p. cm.
Summary: When Gretchen, her twin brother Jason, their little sister Daphne, and a friend find their favorite swimming hole filled with green slime, they learn how the water became polluted and what can be done to clean it up.
ISBN 978-0-88448-327-4 (hardcover : alk. paper)
[1. Water—Pollution—Fiction. 2. Rivers—Fiction.] I. Pastuchiv, Olga, ill. II. Title.
PZ7.C11725Ri 2011
[Fic]—dc22
                                                                                      2010047971

Designed by Geraldine Millham, Westport, Massachusetts
Printed and bound by Versa Press, 1465 Spring Bay Road, East Peoria, IL 60611; January 2013.